Other Books in The Refl

The Miracle of Self-Realization: Reflections on the Spiritual Teachings of Ramana Maharshi by Jeff Carreira

The Miracle of an Open Mind: Reflections on the Philosophy of William James by Jeff Carreira

The Spiritual Implications of Quantum Physics: Reflections on the Nature of Science, Reality and Paradigm Shifts by Jeff Carreira

The Power of Creative Flow: Reflections on Peak Performance, Cultural Transformation, and Spiritual Growth by Jeff Carreira

Inner Peace in a Busy World: Reflections on Meditation and the Journey from Anxiety to Spiritual Freedom by Jeff Carreira

Evolution, Intuition and Reincarnation: Reflections on the spiritual vision of Ralph Waldo Emerson by Jeff Carreira

Free Resources from Jeff Carreira

Life Without Fear: Meditation as an Antidote to Anxiety with Jeff Carreira. Visit lifewithoutfear.online

Secrets of Profound Meditation: Six Spiritual Insights that will Transform Your Life with Jeff Carreira. Visit secretsofprofoundmeditation.com

Foundations of a New Paradigm: A 6-part program designed to shift the way you experience everything with Jeff Carreira. Visit foundationsofanewparadigm.com

THE BATTLE OF SCIENCE AND SPIRITUALITY

REFLECTIONS ON THE HISTORY OF WESTERN PHILOSOPHY

Copyright © 2023 by Jeff Carreira

All rights reserved. Except as permitted under U.S. Copyright Act of 1976, no part of this publication may be reproduced, distributed, or transmitted in any form or by any means, or stored in a database or retrieval system, without the prior written permission of the publisher.

ISBN: 978-1-954642-41-6

Emergence Education
P.O. Box 63767
Philadelphia, PA 19147
EmergenceEducation.com

Cover design by Silvia Rodrigues.
Interior design by Sophie Peirce.

Printed in the United States of America.

THE BATTLE *of* SCIENCE *and* SPIRITUALITY

Reflections on
THE HISTORY OF WESTERN PHILOSOPHY

JEFF CARREIRA

EMERGENCE EDUCATION
Philadelphia, Pennsylvania

Contents

01: Are There Two Kinds of Humans? 1

02: Plato and Aristotle: The Dreamer and
    the Scientist 11

03: The Two Truths of the Middle Ages 23

04: The Heroic Leap from The Middle Ages
    to the Enlightenment 33

05: The Battle of Mind and Matter 43

06: Science, Scientism and the Fate of
    Spirituality in the Western World 53

07: The Triumph of Romanticism 67

08: Existentialism and Pragmatism In
    Defense of Faith 77

09: Postmodernism and Beyond 89

10: Final Thoughts and Conclusions 99

About the Author *105*

JEFF CARREIRA

Science arose from poetry... when times change the two can meet on a higher level as friends.

- JOHANN WOLFGANG VON GOETHE

JEFF CARREIRA

01

Are There Two Kinds of Humans?

*"Science is for those who learn;
poetry, for those who know."*

– PHILIBERT JOSEPH ROUX

JEFF CARREIRA

I HAVE ALWAYS HAD an insatiable curiosity and once I get hooked on something I latch onto it like a dog to a bone. A number of years ago I got hooked on American philosophy. My adventure started with my study of the American Transcendentalist Ralph Waldo Emerson and then the Pragmatist William James. Both of these towering American thinkers are featured in other books in this series. But my adventure didn't stop with the study of American philosophy because I soon realized that in order to understand what people like Emerson, James, Margaret Fuller, and John Dewey were doing, I had to understand the philosophy that was influencing them.

That meant reading philosophers like Hegel and Kant, and poets like Goethe, Coleridge and Wordsworth. Studying German Idealism and Romanticism led me back to Aristotle and Plato and other ancient Greek lovers of wisdom. Then I looked into the scholastics of the Middle Ages and then back to the modern era for a peek at Existentialists like Jean Paul Sarte and Madame de Beauvoir. One thing always led to something else. I read and listened to podcasts and studied for years.

At first I felt like I was just reading words and names. I was lucky if I understood ten percent of what I read. Philosophy books, I decided, are terrible at name dropping, and referencing concepts that they don't explain because they assume you know what they mean or will look them up. I hope I don't do too much of that in this book (but given how many names I have already dropped in just two paragraphs I am a little worried). Unfortunately, a certain amount of name dropping seems to be inevitable, but don't worry, if you don't know a reference or an idea, and I don't explain it, it's because it's not essential.

I remember the first time I was reading a philosophy book and a reference was made to another philosopher and their ideas, and I did know who they were. It was thrilling. I actually do remember this moment. I remember smiling and putting the book down for a second and I felt how what I had previously learned about this author came flooding into my mind and enriched everything that I was currently reading. That was the first time that I realized that somehow the whole story of philosophy fitted together. It was all one story written by many different people over thousands of years.

Gradually the story of how the Western mind had been shaped by philosophical ideas started to form in my mind. I found it thrilling to see how ideas that people came up with shaped how they experienced reality and therefore what was possible for them. A new idea would come along and a whole new world of possibility would open up. The history of philosophy was thrilling to me. And my enthusiasm for the philosopher Alfred North Whitehead's book (oops another name drop!) called *Adventure of Ideas*, is a big part of what I want to share with you.

I am enthusiastic about Western philosophy, but my knowledge is limited, and the space of this small book is even more limited. You can think of this book as offering a crash course in Western philosophy, as long as you keep in mind that it only offers a slight glance. Reading these pages will give you a rapid, but fascinating tour through some of the twists and turns of thought that have led to the development of so many of the ways we currently think and understand the world. It will be a bit like watching a movie on fast-forward, stopping occasionally to see a scene or two in detail. The end result, I hope, will be that you have a better understanding of the history of Western philosophy and deeper insight as to why you think and feel the way you do.

There is one fascinating thread that has weaved its way through the story of Western philosophy that I will be highlighting throughout this book. As I learned more and more about the thinkers that have contributed to the Western understanding of reality, I began to see something. There seem to be two kinds of human beings. Two fundamentally different ways that people see the world, and any given individual seems to be more aligned with one than the other.

The philosopher William James called them tough-minded and tender-minded. There seem to be some people who view the world more through the rationality of the mind and others who see it through the feelings of the heart. People who are thinkers and people who are dreamers, scientists and poets, materialists and spiritualists. Throughout the history of Western thought you see this division show up again and again. Culture, or at least certain segments of it, will be dominated by one side or the other, and then it will flip. The rationalists take over for a while, then the romantics come back.

As I began to see these two opposing forces battling for the control of the Western mind I became fascinated by it. I could see both sides in myself, although I feel that I lean more toward the romantic side of the street. Throughout my life these two sides, my scientist and my dreamer have wrestled for control of my life. In that way, my life is a reflection of the history of Western philosophy as a whole. Scientists and dreamers battling for control. One side holding the reigns of intellectual power for a time, and then the other steps in and takes over.

Many people have noticed the difference and some have come up with explanations as to why there are two fundamental types. Perhaps it is because we have a mind and a heart, thoughts and feelings. Maybe it is because we have an outer experience of the world and an inner experience of mind. You will also discover in this book that this fundamental division in the Western psyche might be caused in part because in the West we have two different definitions for the word truth. Often we speak of something being true because it correlates with facts, but we talk about ourselves being true when we are aligned with our beliefs.

Regardless of the source, this difference seems to be real and in the pages that follow you will see how it has been propagated throughout Western philosophical history. In the modern age, that foundational division is on display in the ongoing battle between science and spirituality. Prior to the Age of Enlightenment, the church was the unquestioned arbiter of truth, but that supremacy was challenged by the new thinking of the age of reason. For hundreds of years the intellectual battle between science and spirituality has been waged. In some ways it seems that science has emerged as the victor, but the struggle is far from over, and after reading the rest of this book, you may begin to see, as I do, that it might never end.

Perhaps this split in the human psyche, regardless of its source, is valuable to us. Maybe questions about fact vs. faith, thought vs. feeling, evidence vs. inner revelation, are supposed to remain open. It might simply be the case that these deep uncertainties about the nature of what is real and true actually help us because they prompt us to explore the source of meaning. I believe that we are not meant to understand everything because not knowing and the desire to know are what keep us moving and alive.

It is fascinating to discover how the way we understand the idea of truth profoundly affects how we think, feel, and perceive. Within Western philosophy we will see how two competing beliefs about where truth is found have contributed to the development of two fundamentally different human temperaments, and how these two temperaments are the root cause of the ongoing conflict between the fact based knowledge of science and the inner revelations of spiritual faith.

"

I returned to poetry as a more precise way to describe the world, more precise than science.

– DAVID WHYTE

JEFF CARREIRA

02

Plato and Aristotle: The Dreamer and the Scientist

"Ideas are the source of all things"

– PLATO

JEFF CARREIRA

THE BATTLE OF SCIENCE AND SPIRITUALITY

ALFRED NORTH WHITEHEAD WAS an English philosopher and mathematician of the early 20th century. He was the mentor of Bertrand Russell, another of the 20th century's most famous philosophers, and together these two esteemed thinkers wrote a massive three volume treatise called *The Principia Mathematica*. This important work attempts to analyze and understand the logic of mathematics so that it can be applied to other forms of thinking. The work of Whitehead and Russell is often seen as part of the birth of analytical philosophy. Analytical philosophy is largely the study of the logic of our truth claims. When we say something is true, what do we mean? How do we know that what we think is true, is actually true?

Analytic philosophy is predicated on the belief that a statement is true when it accurately points to some actuality. A statement is true if it names a truth that exists in reality. In this way of seeing, statements of truth describe facts of reality. This is the correspondence theory of truth, because truth claims are supposed to correspond to facts of reality, and the facts that are described are imagined to exist independent from the statements about them. In other words, what we think and know, has nothing to do with what is true. What's true is true, whether we know it or not. William James' tough-minded personality prefers this kind of truth. They want hard facts, not opinions. They want concrete ideas that can be easily verified in experience.

The central question of this little book is, what is true? And what we'll find, is that in Western philosophy at least, there seems to be two different attitudes that people tend to take toward the idea of truth. One of those attitudes leads to a more romantic, creative, spiritual temperament, and the other leads to a more analytic, rational, scientific temperament. Understanding how these two ways of understanding truth have developed in Western philosophy, helps us understand ourselves and some of the tension we see in our culture, especially the conflict that exists between science and spirituality.

Whitehead is a good choice for starting off our conversation because he was one of the originators of analytic philosophy which represents the rational and scientific attitude toward truth. Later in life, Whitehead developed process philosophy and came out of retirement to chair the school of philosophy at Harvard University in the United States where he expanded on the work of pragmatist philosophers like William James and John Dewey. Whitehead, as you will see, figures prominently in our exploration later on when we talk more about the modern era of philosophy.

RIght now, what I want to say about Whitehead is that he once famously described all of Western philosophy as merely a footnote to Plato. By this he implies that all of philosophy throughout Western history has either been an extension of, or a refutation of, what the great ancient Greek thinker Plato originally thought.

Plato was the intellectual giant who placed the philosophical position known as idealism at the very center of Western thought. Idealism is the belief that reality is ultimately contained in the realm of mind and ideas. According to Plato reality is split into two realms, a physical realm of things and places, and a spiritual realm of thoughts and feelings. For an idealist the ideal realm of thoughts and feelings is more primary, more real, more true, than the physical realm of things and space. The opposite attitude is materialism, in which the physical realm is seen as more real than the ideal. For an idealist, reality is like a dream and all of the things we perceive are like things in a dream; they feel real to our perception, but they are actually being produced by the mind. To a materialist all of the activity of our mind is seen as being produced somehow through the interaction of our biology and the electrical impulses in our brains. Perhaps it is because we have a dual experience of an inner mind and an outer world that two different ways of understanding truth have emerged in our culture. Or perhaps we experience an inner world and an outer world because people like Plato and we will see later, Rene Descartes, thought that way. One of the trippy questions we have to keep asking ourselves as we read this book is *which comes first mind or matter?* Do we think the way we do because of the way reality is, or do we perceive the way reality is because we think the way we do? This is a very important question to ask.

Getting back to Plato, he saw the ideal realm as more primary, more real, and what we experience outside in the physical world as an imperfect reflection of the perfect ideal forms inside us. Many of the great ancient Greek thinkers were fascinated with geometry and mathematics. If we think about it for a minute, we can get a sense of how the attitude of idealism developed through some of the realities of geometry. We can imagine a perfect square whose sides are all exactly equal in length, but in the material world we never find a perfect square. The four sides of any real square whether it be made of wood, metal, or glass, will never be exactly and perfectly equal in length. If you measure carefully enough you will always find a difference. This is why Plato and others thought that mathematics was a very high, if not the highest form of thought. In mathematics we could work with ideal shapes that we can never recreate perfectly on Earth.

If we move away from mathematics and consider ideas of justice, we see the same thing. We can imagine perfect justice, but there are never real lived situations that can be resolved perfectly. There is never perfect justice in practice. Some amount of compromise will always need to be made, and some doubt as to the ultimate fairness that will always exist.

This is the essence of how Plato saw reality. The mind is capable of transcending the limitations of the physical world and in that ideal realm everything exists in its perfect form. The ideal of being human exists in our minds. We can imagine a perfect human being, but no actual human being ever measures up to that perfect state. The same with a horse, or a house, or anything else. We can imagine the ideal, or perfect form of anything, but that ideal can never be fully realized on Earth. Perfection only exists in the realm of thoughts and feelings. Later, in the Middle Ages, this became incorporated into the Medieval synthesis of Christianity as the ideal of heaven, but we aren't there quite yet.

Plato loved the perfection of mind and thought. He was a dreamer who held a certain amount of disdain for the everyday world of our senses. What he truly loved was to allow his awareness to drift off into the unconstrained realm of ideas. Plato had been the student of the great philosopher and teacher Socrates, and in turn Plato's greatest student was Aristotle. Aristotle and Plato held many ideas in common, with at least one important difference between them. Aristotle did not have the same disregard for the physical world that Plato seemed to have. Instead, he was intrigued by the physical world and believed that by closely observing it we could learn nature's secrets and understand how everything worked.

As we said earlier, Plato was the great proponent of the philosophical position known as Idealism. His more earthbound student Aristotle, while perhaps not a materialist, favored the form of inquiry known as Empiricism. Plato, the dreamer, believed that the power of an unconstrained mind and purely creative thinking could uncover the higher truths that would never be found in an imperfect world. Aristotle, with an attitude that today we associate with science, believed that careful observation of the world, imperfect though it may be, is what allows us to see the mysteries of creation at work.

In many ways this distinction between mind and matter, and idealistic and empirical ways of thinking, create a distinction that has propagated through Western culture and often manifests in fundamental divisions that we feel within ourselves, and between people. As we will soon see, throughout the development of Western thought, each of these two positions has attracted proponents and opponents. There have been times and circumstances when one appeared to be superior to the other, and other times in which the former seemed to reign supreme. This same dichotomy shows again and again in opposing conceptions that stubbornly resist resolution. Some of these include the split between heart and mind, Heaven and Earth, body and soul.

As we continue our journey you will see that this underlying sense of division is part of your experience of reality and central to your thinking. If we take the split between mind and matter as an example, we find that it seems so obviously real to us that it is hard to imagine that it might not be true. But perhaps it is not, maybe it is a distinction that only exists because we have learned to see it that way. We have inside us a number of different attitudes towards the truth, and corresponding ways to determine what the truth is. Those of us with a more creative and romantic disposition tend to favor one way, those with a more scientific and rational disposition tend to favor the other. We see an example of this fundamental distinction in the attitudes of Plato and Aristotle. In the next chapter we will explore how two dramatically different conceptions of truth developed in the Middle Ages during a fascinating period of time during which the philosophical wisdom of the Greeks mixed with the religious ideas of Christianity.

JEFF CARREIRA

"

Educating the mind without educating the heart is no education at all

- ARISTOTLE

JEFF CARREIRA

03

The Two Truths of the Middle Ages

"I am the way and the truth and the life."

~ JESUS

JEFF CARREIRA

THE BATTLE OF SCIENCE AND SPIRITUALITY

During the trial of Jesus we are told that Jesus was offered a plea deal. If he was willing to deny that he was the king of the Jews, then his life would be spared. All he had to do was make one simple statement and he could live, but he didn't. He refused to deny the statement and accepted the punishment of a brutal and painful death by crucifixion. He certainly had all the motive in the world to make a simple statement of refutation and avoid such a dreadful demise, and yet he chose not to.

Later in the 16th century during another famous trial the scientist Galileo Galilei was offered a similar proposal. He was also facing a death sentence and all he needed to do to avoid it was retract his declaration that the Earth revolved around the Sun. That's it. Just say you're wrong, and you can avoid a horrible end. Unlike Jesus, Galileo took the deal. He retracted his findings and said he was wrong. He avoided punishment, but then, as he left the courtroom he supposedly muttered, "And yet it moves." In other words, you can force me to say anything you want about the Earth and the Sun, but the truth about them will remain unchanged regardless of what you make me say.

Is this a case of two different levels of integrity? Or is something else going on here? Did Jesus care more about his mission than Galileo did about his? Why was Jesus compelled to take a stand for what he considered the truth even when faced with his own death, while Galileo was content to deny what he knew and live on? It might not be because one man had more integrity than the other; it might be because they were each operating from a different understanding of truth. And as we began to explain at the end of the last chapter, both of those ways of seeing truth are part of the canon of Western philosophy and therefore part of the way that you and I experience truth today.

Central to the "good news" that Jesus and his disciples were spreading was the announcement that the kingdom of Heaven was here. Jesus was preaching a doctrine that promised the possibility of Heaven on Earth. Heaven was not just a transcendent realm of perfection that could only be attained after your physical existence on Earth was done; Heaven is here, or at least it can be, if we take a stand for that possibility. If we live as if Heaven is here, we will bring Heaven to Earth. The truth of Heaven on Earth depends on us. It depends on our willingness to live as if that were true and stand unwaveringly in that conviction.

In this case, truth is not just a fact about a reality

that exists entirely independently from us. Truth is dependent on us and the stand we take for it. How we are and what we do makes things true. Truth is not an inherent quality of things; truth is a stand that must be taken. Because Jesus was acting in accordance with this understanding of truth he could not deny the accusations against him. The possibility of Heaven on Earth that he was committed to, depended on his willingness to stand for it.

Galileo, on the other hand, was a pioneer of science and science was heavily influenced by the Greek philosophical tradition. In the Greek tradition, as we saw in the last chapter, truth exists in an ideal realm beyond this world. What is true, is true beyond our reach. We can't touch it and nothing we do can change the fact of it. With this conception of truth there is no harm at all in Galileo denying that the Earth rotates around the Sun. That will be the case no matter what he says anyway. It was true before anyone ever knew it, and it will continue to be true even if everyone forgets. What is true is true and nothing we do matters. There is no reason to take a stand for the truth here on Earth because what is true will be true whether we stand for it or not. In this case, truth is not a stand that we take; it is a fact that exists completely independent of what we do. This conception of truth left Galileo's conscience completely clean even as he denied what he knew was true.

These two different conceptions of truth are deeply

embedded in the way you and I have been taught to think. They are a deep part of our cultural and philosophical heritage. We commonly speak about 'being true' to our word. That means that our actions are aligned with what we believe is true. Being true in this sense means that when we believe something, our actions demonstrate that belief. Of course, we also speak about things being true, meaning that they are facts about reality. In this sense truth means that the statement we make about what is true corresponds to actualities that exist in reality independent of us. These things are 'true' whether we happen to believe in them or not. In the first case truth is a stand that we take, in the second case it is a fact that is valid regardless of us.

After the fall of Rome, the Western world fell into what is called the Middle Ages. During that time, life for most people seems to have been characterized by hopelessness and brutality. It was a time of widespread war, poverty, lack of education, and a loss of the richness of culture. The two major intellectual influences on those times were the doctrines of the Christian Church and the writings of some of the ancient Greek thinkers, especially those of Aristotle.

During this dark time in Western history a worldview gradually developed that combined Christian ideals and Greek thought. This world view is often known today as the Medieval Synthesis. It was the worldview of what is sometimes seen as the classical period of Western culture and it held the Western world together during its bleakest period. Eventually the ideas of the classical period were challenged and ultimately overthrown by the Scientific Revolution of the Age of Enlightenment. In the next chapter we will see how the intellect of the Western world moved from the Middle Ages into the Enlightenment. In this transition the classical era stepped aside to make way for the modern world.

JEFF CARREIRA

And yet it moves.

– GALILEO

JEFF CARREIRA

04

The Heroic Leap from The Middle Ages to the Enlightenment

"Whatever appears as a motion of the sun is really due rather to the motion of the earth"

~ NICOLAUS COPERNICUS

JEFF CARREIRA

THE BATTLE OF SCIENCE AND SPIRITUALITY

THE LEAP THAT OCCURRED in human consciousness that took the Western world from the Middle Ages through The Age of Enlightenment and then into the modern world, is nothing short of miraculous. Try to imagine being alive during the Middle Ages. The vast majority of people were almost entirely uneducated in the modern sense. Sometimes I try to think about what it would have been like to have been a serf in the ninth century. I imagine that most serfs had very little cognitive activity that we would recognize as thought. Probably most of a serf's inner experience would be composed of strong and often conflicting emotions. Feelings of fear, desire, ecstasy, and rage, might alternate rapidly through their system. Of course, I have no way of knowing, but certainly the inner experience of a medieval serf would be dramatically different from ours today.

As an uneducated serf, your entire understanding of the world, what little you had, would have been made up out of a combination of Christian doctrine, local folklore and superstition, with for the lucky few, maybe a sprinkle of Aristotelian rationality. Of course there were educated people at the time, mostly within the church among the nobility, and among those who practiced the old religions, but they were relatively few. From our present day point of view, it would be as if no one had any understanding at all.

Given the difficulties and challenges of the time, the experience of the average person was undoubtedly dominated by fear. He or she lived in a world with danger on all sides. Those who were strong would take what they wanted and might easily kill those who resisted. Your greatest source of hope in this bleak landscape came from the church that preached about an all-powerful king who lived in a heavenly realm and who promised to reward the downtrodden in the afterlife. You would have very little sense of causality and so the world you lived in would appear chaotic and full of random occurrences. What little understanding you did have would likely be explained by a logical form that we wouldn't even recognize as rational today.

When something as devastating as the bubonic plague swept through Europe, it killed somewhere between 25% and 60% of the population. There was no understanding of infectious disease and so people tried to create the best understanding they could to explain why so many people were dying. The worldview of the time would tend to convince you that either evil spirits were attacking the world, or God was punishing humanity for some regression into sin. You would have no way to imagine that your culture's lack of personal hygiene could account for the rampant spread of the illness.

The world of the Middle Ages would have been a radically unpredictable one, full of brutal contradictions. The same God who benevolently provided the Earth under your feet, air to breathe, and food to eat, also left you riddled with disease and ill health, and periodically whipped up all sorts of natural and human disasters. There was no clear way to act that would accurately allow you to predict or control the future, and the God you worshiped and feared was an impossible mix of generosity and cruelty.

In these bleak times there were always those who held to higher ideals and who worked hard to incubate knowledge in the hopes that the world of the future would be ready to make use of it. I once saw a documentary about the history of the Middle Ages. It wasn't the best documentary I've ever seen, but I was engrossed. It was a seven part series and each episode covered a span of about a century. What I noticed as I watched was that every fifty to a hundred years someone would get the idea that they could inspire people to live in a more harmonized and civilized way. To be clear, these periodic leaders were just about as brutal as anyone else, but they had a higher vision for humanity and they tried to enact it. Invariably they failed, but then some decades later, someone else would try. At one point after watching about five episodes I started weeping because I saw how the human spirit cannot be held down for long. No matter how bad things get, there is something in humanity that will envision a better possibility and struggle for it. Even though time after time people failed, someone else would always get up and try again.

Many of the most highly educated individuals of the Middle Ages were Christian monks who were heroically busy in monasteries studying texts from ancient Greece and synthesizing the ideas they found with Christian doctrine. The brilliant theologian Thomas Aquinas is often credited with creating the most comprehensive synthesis of Medieval wisdom. This masterful achievement created a worldview that held the Western world together during some of its darkest hours.

The worldview of the Medieval synthesis was the dominant framework for thinking until a Polish born astronomer named Nicholas Copernicus rocked its foundation by pulling out one of the most significant pillars holding it up. When Copernicus showed convincingly that the Earth revolved around the Sun and not the Sun around the Earth, it was the beginning of the end of the classical worldview. Copernicus' discovery challenged one of the central characteristics of Christian thought – the centrality of humanity's place in the universe and thus humanity's favor in the eyes of God. Over the next few centuries this crack in the cosmic order would multiply and grow into one of the greatest intellectual revolutions in human history. Soon the German born astronomer Johannes Kepler would show that the planets revolved around Earth according to simple mathematical relationships, and still later the Englishman Sir Isaac Newton explained the motion of the planets using his simple and elegant conception of gravity.

The Enlightenment changed the universe. Suddenly it was clear that the universe wasn't an unknowable place to be feared. It was an organized mechanism consisting of different parts that acted in accordance with natural laws that could be discovered and understood. The universe was knowable and we had the ability to understand and perhaps control it using the power of reason. That was the simple, magnificent, and transformative central principle at the heart of the Enlightenment.

The universe that emerged out of the Enlightenment resembled a "clockwork." It was not a universe dominated by Gods and spirits. Everything worked according to natural laws that held true always and everywhere. By understanding the rules by which everything worked we could predict and to a large extent control the future of events. This realization led to an explosive utopian impulse within the hearts and minds of many Enlightenment thinkers. It was now clear that we could find the keys that would allow us to perfect the world.

The American founders, Benjamin Franklin, Thomas Jefferson, James Madison and others, were Enlightenment thinkers who held a vision for a perfect form of governance that would work in accord with the universal laws that they felt guaranteed life, liberty and the pursuit of happiness as the inherent rights granted by existence. They, along with many others, risked everything including their own lives in order to create this new world.

The Enlightenment was a magnificent leap forward for the Western world and for a time it seemed like final solutions to every problem would soon be discovered and the vast unknowable mysteries of the universe would be entirely illuminated. Total and complete understanding of the workings of the universe was the promise of The Enlightenment and it was a promise that has dominated Western thought, for better and worse, right up to the present today.

"

*Following the light of the sun,
we left the Old World.*

~ CHRISTOPHER COLUMBUS

JEFF CARREIRA

05

The Battle of Mind and Matter

"The good thing about science is that it's true whether or not you believe in it."

~ NEIL DEGRASSE TYSON

JEFF CARREIRA

As the Western world moved beyond the mind of the Middle Ages and approached the dawning of enlightenment, the stage was set for an intellectual struggle to begin that would decide who would control the Western mind. It was a dilemma that was inevitable, because the world now simultaneously held two competing worldviews. The God centered worldview of the classical world, and the reason centered worldview of the enlightenment. The battle lines that were being drawn pitted faith against reason and spirituality against science

As we saw in the second chapter of this book, Plato and the Greeks were idealists. They believed that ideas were real and what occurred in the world would always be imperfect. In this framework, truth is verified in reference to universal ideals that exist outside of the world and remain unchanged throughout time regardless of what any of us does. The natural laws of science in many ways fit neatly into the Greek notion of Idealism. The law of gravity for instance, is "true" whether you believe it or not. We can deny the "truth" of it, but we will still fall when we jump out of a tree.

There is also a great deal of idealism at the center of the Judeo-Christian tradition. Heaven is, after all, an ideal realm of ultimate truth. At the same time, in this religious tradition what we do on Earth also matters. The truth cannot simply be true out in some ideal realm. We are charged to take a stand for the truth right here in the world. The phrase 'on Earth as it is in Heaven' asks us to make the universal laws of heaven equally true on Earth. Truth is not just a fact, it is a stand that must be upheld.

As we have already begun to discuss, these two opposing conceptions of truth can be seen reflected in the two opposing schools of thought that developed at the onset of the age of reason. These two ways of looking at the world are known as Rationalism and Empiricism.

A rationalistic thinker puts their faith in reason, in the mind, and in thoughtful analysis. They mistrust sense experience because they recognize that our senses can be deceived. A straight stick placed in water looks bent for example. We see the sun move through the sky and assume it must revolve around Earth, that is until we realize that Earth is rotating regardless of what the sun appears to be doing in the sky. To the rationalist, truth lies in the realm of ideas and reason. The discerning power of the mind should be trusted over the impressions of our senses.

Empiricists, on the other hand, put their faith in our sense experience of the world because they believe our ideas and opinions are often clouded by our preconceived attitudes and prejudices. Ten people can experience the same situation and draw ten different sets of opinions about what happened. If we want to believe something, we can always come up with a way to rationalise that belief. For these reasons empiricists mistrust ideas that are not grounded in the sensual evidence of observation because without that grounding, ideas too easily fly into fantasy. Empiricists want proof of the truth, and they want proof that can be verified in the actuality of experience. The conflict between empiricism and rationalism can easily be found today in the tension that exists between science and spirituality. This is a tension that we will continue to explore over the next few chapters because it represents one of the central underlying questions about truth and reality that any thoughtful person must grapple with.

Science is an empirical pursuit. It uses the scientific method to build an understanding of the world based on observable and verifiable facts. The scientific method involves hypothesizing and developing a theory that might explain something. The scientist then designs and conducts experiments to prove, or disprove, their theory. Only information that is obtained through experimental observation is considered scientifically valid. Only ideas that have been tested and proven are believed worthy of being considered true, and even though only provisionally because all scientifically proven theories are held open to continued examination. Nothing in science is ever considered to be absolutely true forever. Every theory can be replaced by a better theory and a new understanding.

Spirituality is a more rationalistic pursuit that believes in the validity of truths that are revealed to us only in our inner experience. These inner truths of revelation are considered worthy of belief even if they cannot be validated through experimentation. Spirituality has as its goal the moral transformation of the individual so that they are better able to act in accordance with the revealed truths of their inner vision. These revealed truths are seen as being self-evident and therefore beyond the necessity of experimental tests to "prove" them. Spirituality believes in the power of faith.

Over the past few centuries science has increasingly become the dominant framework for thinking in the Western world. It has largely replaced religion as the ultimate arbiter of truth. We live in a scientific age and we all subscribe to the tenets of the scientific method more than we might realize, even if we consider ourselves to be spiritual or religious. We all tend to want to prove the validity of anything we choose to believe and we will generally trust ideas that we think have been scientifically proven, over those that appear to be merely speculative.

In an intellectual atmosphere dominated by

science, many have asked *what is the significance of spirituality in the modern age?* Is spirituality anything more than a set of personal beliefs that make us feel better and help us cope with life's inevitable challenges? As we contemplate the value of spirituality in the modern world we inevitably run headlong into the questions of Empiricism vs. Rationalism and Materialism vs. Idealism, which can be summed up neatly and stated as: Are we intelligent matter – stuff that got smart? – or are we an incarnate spirit – smarts that grew stuff around it?

Many great religious traditions are idealistic in nature. They believe that some form of higher mind is the source of existence. These traditions tend to see us as spiritual beings who became flesh. First there was God or pure spirit, and then from God came us. At our deepest level we are also a spiritual being with a soul that exists beyond the physical world. Our more recent scientific understanding of reality has led many people to believe the opposite, that we are fundamentally organic beings composed of material stuff that eventually evolved and became alive and intelligent.

If we are essentially a spiritual being - a soul - that has taken physical form, it means that at least some part of us exists separate from, and outside of, the physical universe. It implies that the source of our intelligence exists outside the natural world and acts upon the world from there. In this view, we are above the laws of nature and therefore uniquely autonomous and responsible as the source of our own action in the universe. We have an inherent moral responsibility because of the spiritual and otherworldly source of our being.

If, on the other hand, our consciousness is a biproduct of complex interactions of physical matter and natural forces, then we are a product of nature herself and subject to her laws. Our actions and thoughts are not sourced from an outside reference point, they are a necessary consequence of an intricate chain of cause and effect. Our actions result from natural interactions the same way that the movement of a tree results from the blowing wind. This view fundamentally calls into question the idea of human freewill, autonomy and ultimately any notion of responsibility.

Can you see the dilemma? There are aspects of both views that probably appeal to you. On the one hand you might feel that there is a deeper part of us that is not simply the result of material interactions, on the other hand you might also believe that we are a part of nature and not separate from it. In many ways this is the conflict that has split the modern world in half. Lining up on one side are those that believe in the spiritual world of religious tradition, and on the other side are those who are champions of rationalistic logic and the knowledge of science. In reality, most of us are a mixture of the two and so we must struggle to rectify opposing parts of ourselves.

> *We are spiritual beings having a human experience.*
>
> – PIERRE TEILHARD DE CHARDIN

Science, Scientism and the Fate of Spirituality in the Western World

"Postulates are based on assumption and adhered to by faith."

– ISAAC ASIMOV

IN THE LAST CHAPTER we witnessed the transition of the Western world from the Middle Ages through the Enlightenment and into the modern age. What we found emerging were new variations of the intellectual conflict between the ideal and the real. Same battle, different battleground.

Does God exist? Do we have freewill? These are just two forms of one of the biggest and most challenging underlying questions of philosophy. The premise of this book is that most of the questions that arise in philosophy stem from one really BIG question about the nature and foundation of reality. As we have been exploring, this really big question asks if reality is rooted in some transcendent realm that can only be experienced in our minds, or is reality first from a physical world that is experienced through the senses of physical beings. Are we sourced from a transcendent spirit, or are we merely a byproduct of material interactions?

Big questions like these are part of the branch of

philosophy known as ontology or metaphysics. In these disciplines philosophers try to determine what is real. Throughout the history of philosophy, metaphysical questions have been debated endlessly without resolution. The challenge of determining what is true about reality gives rise to the need for a branch of philosophy called epistemology which concerns itself with questions about how we can determine when something is true. Where metaphysics is concerned with determining what is true, epistemology is busy exploring how we can be sure about what we think. In short, epistemology is the study of how we know things. We can see in our own experience the essence of these two branches of philosophy when someone asserts that something unusual is true and we immediately feel compelled to ask, how do you know that? Any legitimate search for truth must include both speculations about what is real, and also an examination of how we know it.

Today the scientific method is the most prevalent epistemological orientation in the modern world. As we have said already, science tells us that truth is found in experimental results. We all tend to be skeptical and feel that we must test and prove our ideas before we can believe they are true. But what happens when more than one theory can fit the same experimental results? How do we know which of them is true then?

Another epistemological idea that has been incorporated into the way we think was developed by

William of Ockham, a monk in the Middle Ages. This theory is often called Ockham's razor and it basically states that whenever there are two theories that can equally explain the same phenomena, the one that requires the least number of assumptions is the one we should accept as true. In other words, when looking for truth we should always look for the simplest explanation possible.

It is this epistemological attitude that makes the scientific mind uncomfortable with spirituality, because it sees belief in the existence of God, or divinity, or supernatural beings of any type, as huge assumptions to make. After all, why should we assume the existence of a mysterious intelligence if we don't have to? As humanity emerged from the Middle Ages a battle began to be waged between the leaders of the church and the new intellectuals of the Enlightenment. That struggle continues today even though it may appear to many that the debate is over and that science won.

Over the course of the 19th and 20th centuries the discoveries of science and the rapid improvements to human life they created were simply too much for religion to contend with. There seemed to be mounting evidence that science provided superior access to truth and the battle for the Western mind appeared to be nearly won. In fact, many people still feel that the debate is over, and that science has proven itself the superior way of knowing. To these people, it is obvious that the experimental method of science is the supreme means for discovering what is true. For others the debate is far from over.

Certainly science has proven magnificently effective in helping us understand, and to a large extent, control the physical world. We have all benefited tremendously from the triumphs of science, and yet we find ourselves perched on the edge of global destruction, facing a host of problems that science seems incapable of solving and in some cases has even had a hand in creating.

Some wonder if we have reached the limits of science, and some feel that the scientific paradigm is itself somehow responsible for some of our global problems. I often say that the paradigm we are in needs to change, but I would like to be clear that I do not see our current challenges being caused by the epistemology of the scientific method. I was originally trained as a scientist and have deep respect for the epistemology of science. The challenge we face is not with the scientific method, or science per se. As I see it, one of the biggest problems with the current paradigm has to do with an attitude about science that has developed due to the enormous success that science demonstrated. This faulty epistemology, which is not a reflection of the scientific method, is called scientism, and we should make a sharp distinction between science and scientism.

As we have been discussing, science is essentially a method of inquiry and the body of knowledge acquired by that method. The scientific method involves hypothesizing, experimenting, observing and drawing conclusions based on what you see and measure. It is also central to the philosophy of science that no theory is ever considered to represent the final truth. All theories, regardless of how rigorous they seem, continue to be subject to further scrutiny and reexamination, and it is assumed that every theory will potentially be proven false by a more comprehensive theory later.

Scientism, on the other hand, is the belief that the methods of science are superior to any and all other methods of determining truth, and that proven scientific theories represent the real and final truth of the way things are. In short, science tentatively accepts the reality of things that have been observed to be true, while scientism believes that all of the things that science accepts as true actually are. In the modern world we have all been conditioned to believe that the things science believes actually represent the way things are, not just our best guess so far, but the actuality of things.

Most of us are guilty of scientism to some extent. We hear about things that have been scientifically proven and we tend to assume that means they're true. We didn't do the experiments. We didn't look up the results. We didn't analyze the data. We simply assumed that if science proved it, it must be true. There are often good reasons to feel justified in this assumption, and as a former scientist myself, I would never want to promote a fundamental disbelief in the findings of science. At the same time, anyone who has ever done scientific work knows that experimental evidence is always interpreted, and interpretations are always fallible.

To the extent that scientism dominates or even influences our thinking, it would be good to question our faith in science to be sure that we are using our own good judgement when we make decisions about what to believe. Beyond the need to more deeply question what we choose to believe, there is an attitude behind both science and scientism that we would do well to question.

Long after the Enlightenment a group of philosophers and academics gathered in Vienna, Austria in the wake of World War II. They wanted to understand how the world had descended into the chaos and violence brought on by the rise of the Third Reich of the Nazis. Calling themselves the Vienna Circle they created the outlines of a scientific worldview. They believed that by extending the logic and methods of science to the rest of life we could avoid the temptation to believe in dangerous false ideas in the future. The circle of thinkers outlined a philosophical position known as Logical Positivism which asserted that all truth should be verified through observation. Only analytically provable claims should be accepted as true. You and I and almost anyone likely to read this book is probably at heart a logical positivist without even knowing it. This is because most of us tend to assume that something is only true if and when there is conclusive evidence proving it. In other words we don't tend to take much on faith!

But, why do we think that conclusive evidence is the best way to tell what is true?

For a couple of centuries after the Enlightenment the evidence-based logic of science seemed to have won a decisive victory over all other ways of knowing. The wisdom of faith and the revelations of religion in many quarters were on the run, often scrambling to justify their existence. Eventually however, considerations were raised that challenged the scientific worldview and its claim to epistemological supremacy.

Modern technologies and conveniences seemed to lead to environmental degradation, industrialization seemed to be destroying the quality of many people's lives, and the inequality between the most wealthy and the poorest grew. Some began to believe that we had lost something very valuable when we lost our ability to believe in anything that we could not see with our eyes.

Perhaps mystery and uncertainty were part of the fabric of the universe and should not be cast aside. It might turn out that the ultimate truth is always a matter of faith, and exercising what the great American philosopher and psychologist William James called "our will to believe" was an essential part of human life. In fact, in the end even the position of Logical Positivism is a matter of faith, because the idea that conclusive evidence is the best way to know what is true is itself taken on faith.

It is interesting to ask yourself, am I a modern thinker? I believe that you are and here are five habits that describe modern thought.

1. You are suspicious of emotionalism and prefer to make decisions with a cool head.
2. You know that if you are emotionally involved with a situation, it is difficult if not impossible to see it objectively and therefore accurately.
3. You believe that if all personal biases, prejudices, and limitations were removed everyone would see the same truth of a given situation.
4. You believe that the universe operates according to natural laws.
5. You believe that the human mind, given enough time and energy, can figure out just about anything.

If these statements apply to you then you are a modern thinker.

As a modernist you believe in the power of mind, especially when it is liberated from undue emotional influences and prejudices. You see a universe that is ordered and operates according to natural law, and that it can therefore be understood. You are suspicious of strong adherence to inner revelation and prefer to seek for evidence to support what you believe. These attributes are a large part of our intellectual heritage. They guide a great deal of our thinking and have served our species in many ways exceedingly well.

In the next chapter we will see how some of the limitations of this way of thinking led to another revolution in Western philosophy - the Romantic revolution.

JEFF CARREIRA

"

I think it's much more interesting to live not knowing than to have answers which might be wrong.

~ RICHARD P. FEYNMAN

JEFF CARREIRA

07

The Triumph of Romanticism

"All the knowledge I possess everyone else can acquire, but my heart is all my own."

- JOHANN WOLFGANG VON GOETHE

JEFF CARREIRA

As we've already discussed, there was a time during the 18th century in which to many it seemed that the rationality of the Enlightenment had triumphed as supreme over all other ways of knowing. Those devoted to this rationalistic way of thinking rejoiced that soon the laws that operated behind all the mysteries of the universe would soon be uncovered and understood and humanity would create the future it truly wanted.

Romanticism as an intellectual revolution was born out of a sense of disillusionment with the shortcomings of Enlightenment thinking just as the Enlightenment had emerged out of the shortcomings of the worldview of the Middle Ages. Philosophically the Enlightenment tendency to believe only what could be proven, had early on started to back itself into a corner.

The philosopher David Hume took the necessity of evidence to its ultimately skeptical end and came to believe that we can know nothing at all, because all we will ever have access to is our sense perceptions and there is no way to know if those perceptions correspond to any 'real' world beyond them. Whether reality is a physical world of time and space or a transcendent realm of spirit, there is simply no way to be sure about any aspect of it. In fact, there was no way to know if there was any reality at all outside of the perceptions of our senses. Hume fell into such despair over this profoundly skeptical position that he was known to frequent public backgammon games in order to get his mind off of the radical insecurity of the human predicament.

Another, perhaps more tangible shortcoming of the Enlightenment was the failure of the French revolution. What started as a revolt against tyranny with the aim of putting in place a government created according to the highest principles of enlightened thought, turned into a bloodbath that seemed to showcase the lowest aspects of human character. What did it mean? The French revolution was meant to bring peace and harmony and a new order to governance, what had gone wrong?

Some intellectuals, artists, and writers throughout Europe began to feel that in our rapid embrace of the power of knowledge we had missed something very important. The reason why the French revolution and the advances of rapid industrialization hadn't turned out to be the panaceas that so many hoped they would be, was because there was something about the way people were thinking that was suffocating the human spirit and squeezing passion and morality out of existence. The German philosopher Immaneul Kant created a new vision of reality. He didn't subscribe to the view that behind everything that happened was a set of universal laws that could be discovered, understood and controlled. Instead he envisioned a universe that grew like a tree and he further saw that human choices and acts of human will were a part of how that growth occurred. The seeds of thought that Kant planted blossomed later into a philosophical movement called German Idealism and it created the intellectual underpinning of what would later be called Romanticism.

Where the Enlightenment saw a mechanical universe run by fixed laws, the new thinking of Romanticism saw a universe that was organic and grew in accordance with acts of will. The Enlightenment had held human reason and rationality in the highest regard. The Romantics elevated human will and creative freedom to a stature equal to or even above reason.

The Romantics were skeptical of science. Frankenstein, the great romantic novel by Mary Shelly, is the well-known story of a scientist who pieces together a human form and brings it to life by harnessing the power of lightning. His creation, a hideous being created from randomly acquired body parts from graveyards, is not the beautiful achievement that Dr. Frankenstein imagined. It seems that life cannot be controlled by science and in the end Dr. Frankenstein's monster destroys him and those around him.

The story leaves us wondering who the monster really is? the creature or the man who created it?

The Romantics felt that the Enlightenment notion that the universe was knowable and controllable was naive. The universe was infinite, mysterious and ultimately unknowable. We do not exist as beings separate from the universe. We do not possess a wisdom that is higher than nature. We are part of the universe, part of the natural world, not above it. We do not discover the mysteries of creation by standing back and thinking about things. We unleash our true creative potential by giving ourselves to our deepest yearnings until we become one with a cosmic process of creation.

For the Romantics, the highest human value was not rationality, rather it was authenticity, moral integrity and passion. The Romantics were among the first to value these human attributes for their own sake regardless of what they were aimed at. For instance, a Christian in the Middle Ages would never admire the zeal a pagan showed for their heathen faith. The Christian would simply see the zealous pagan as that much more wrong and misguided and therefore more dangerous. The Enlightened thinker didn't admire the monk's passionate love for God, instead the monk seemed all the more foolish for it. The Romantics, on the other hand, admired the passion of their enemies. To die for one's ideals was one of the highest goods and it is good no matter what the ideal. Even if someone is willing to die for something you abhor, you can hate the reason, but still admire the passionate commitment. Because of this, in romantic literature you find stories in which the antagonist, who is clearly the character you like the least, is also the character you admire the most.

If the Enlightenment thinkers had felt shackled by the superstition of the Middle Ages, the Romantic thinkers felt straightjacketed by the rigid adherence to the natural law of the Enlightenment. The Romantics loved to break rules, to snub laws and live as unconventionally as possible. They were original in dress, in lifestyle, and in thinking. As poets, playwrights and novelists they broke literary styles. The great romantic composers, perhaps Beethoven greatest of all, were notorious for breaking musical convention.

In Germany, the writings of Kant, and those that he inspired, Fichte, Schelling, and Hegel, and perhaps most importantly of all the playwright Johann Wolfgang von Goethe, set the stage for the Romantic revolution. This revolution would simultaneously erupt in the English poets Byron, Shelly, Blake, Coleridge and Wordsworth. And later find its way into American thought through the writings of Concord Transcendentalists like Ralph Waldo Emerson, Henry David Thoreau, Margaret Fuller, Nathaniel Hawthorne, and others.

It is interesting to ask yourself, am I a Romantic?

I believe that you are and here are five habits that describe modern day romantics.

1. You recognize that human beings cannot disconnect themselves from the natural world.
2. You have a sense of style and accept that some choices are made purely based on aesthetic considerations.
3. You see yourself as a unique individual who cannot be replaced by any other.
4. When you eat at a restaurant you think about what you want to eat and don't assume it will be the same as yesterday.
5. You sometimes go window-shopping without needing anything, just to look at things you could buy if you wanted to.

If these statements apply to you then you're a Romantic.

As a Romantic you possess a consciousness that appreciates both the indivisible wholeness of life and the irreplaceable uniqueness of the individual. You don't believe that the universe is totally understandable. You accept that life is infinite and some aspects of it will always remain beyond our mind's ability to understand. You are suspicious of the use of knowledge and power to manipulate and control circumstances. Instead you prefer to seek for a more easeful flow that puts you in harmony with the life process and allows a deeper creative impulse to work through you. Most of us in the world today are both modernists and romantics.

"

We of the craft are all crazy.

~ LORD BYRON

JEFF CARREIRA

08

Existentialism and Pragmatism In Defense of Faith

*"There are two ways to be fooled.
One is to believe what isn't true;
the other is to refuse to believe what is true."*

– SOREN KIERKEGAARD

JEFF CARREIRA

THE BATTLE OF SCIENCE AND SPIRITUALITY

During the late 19th and early 20th centuries two philosophical perspectives emerged to challenge the supremacy of rationalism and the attitude of logical positivism. One of these was the philosophy of Existentialism developed in continental Europe, the other was Pragmatism developed in America. In this chapter we'll take a look at each of these philosophies and what they were responding to and how they attempted to tip the balance of power toward the romantic impulse by creating what you could think of as more rational forms of Romanticism.

William Barrett introduced the philosophy of Existentialism to an American audience with the publication of his book *Irrational Man*. In the book he is clear to state that Existentialism is strictly a continental European philosophical movement. Although he adds that "of all non-European philosophers William James probably best deserves to be labeled an Existentialist." In fact Barrett goes so far as to claim that it would be more accurate to call James an Existentialist than a Pragmatist.

As the nineteenth century rolled into the twentieth, the Existentialists began to realize that the modern world was heading toward an impending tragedy. The Enlightenment and the subsequent triumph of science and rationalism had eroded the stronghold of faith that had held humanity together for centuries. The dogma of the Christian church simply did not seem capable of surviving the challenge presented by the new understanding of the universe that science had brought. "God is dead," Friedrich Nietzsche announced.

The Romantics had realized that in our rush to rid ourselves of superstition we had lost something essential to human life. We had left a void in the human heart where faith and imagination once lived. The Romantics looked nostalgically back to the Middle Ages and beyond, longing for the sense of awe and wonder that the Christian world had contained.

Through their poetry, prose and music they attempted to re-enchant the world with spirit and a sense of mystery.

A century later the Existentialists realized that human consciousness had become increasingly dominated by rationalism and materialism and that humankind was rapidly losing the ability to have faith in anything that could not be seen or acquired. Humanity appeared to be heading down the road of nihilism.

Some Existentialist thinkers, most notably Soren Kierkegaard, Leo Tolstoy and Martin Buber searched for a new footing upon which faith could reestablish itself in the modern world.

Kierkegaard taught that we must find a mature faith to replace the simplistic blind faith of the past. The maturing of faith, he claimed, could not come from proof; it had to be won by consciously taking the risk to believe in something higher without proof. He felt that any attempt to use reason and rationality to prove the existence of God would do a disservice to the power of faith. Belief in God is powerful because it involves risk and demands that we make a leap of faith. If we were to prove God's existence through evidence, then believing would not involve faith, and faith is the true source of spiritual power. God without faith would have no power in the world.

Existentialism and Pragmatism are not sets of beliefs about what is true; they are attitudes toward life and ways of inquiring into what's true. Different existentialists came to different conclusions about what is true and some, unlike Kierkegaard, felt that our maturity as a species involved giving up all beliefs, including those of God, that offered only a false sense of security in an uncertain universe. The idea was that humanity had outgrown faith and was maturing to the point that we were willing and able to face the fact that our fate is terrifyingly unknown. Human beings did not need to update the spirituality of the Middle

Ages; we needed to find new ways to bring meaning, truth and goodness into the mysterious and ultimately unknowable universe we found ourselves in. We had to find a new source of faith, but not one that rests on mythical beliefs about a creator God or a heavenly realm in the clouds. The existentialists were calling for a stark confrontation with the fact that humanity had grown beyond the old sources of security and must now face the emptiness that lies at the heart of human existence. Human beings needed to find an alternative to faith that could fuel the human spirit and propel it forward into an unknown future. It was time to walk into the unknown armed with only our creative powers and build a new future for us all. The American Pragmatists developed a philosophy that offered an alternative understanding of truth.

William James was one of the American founders of the philosophy called Pragmatism. Pragmatism is largely epistemological because it offers a new way for us to know what is true and rests on a different conception of truth altogether. Rather than the truth of an idea being seen as a characteristic of the idea itself, a pragmatist believes that ideas reveal themselves to be true only when acted upon. Those ideas that are true will improve life when acted upon, those that are false will be shown to degrade life in some way. The truth of an idea is not a quality of the idea itself, it is determined by imagining how believing in that idea will affect human life. Rather than worrying about whether or not an idea is true in the sense that it corresponds to something in reality, we should be concerned about how that idea will change our lives if we believe in it. If believing it makes a positive difference in life, that is as true as you can get. This was indeed a pragmatic approach to truth.

James was a scientist and an academic philosopher who, like Kierkegaard, was deeply concerned with questions of spirituality and religion. He risked his reputation by using pragmatic logic to argue against a scientific world that no longer had room for faith. In the face of intense opposition James vehemently defended our right to believe in things without evidence.

Essentially, James was challenging the philosophical position of Logical Positivism. He was very directly saying that we should not only believe in things when we have conclusive evidence that prove them to be true, we must also be able to believe in things for which we can never obtain proof. He even wondered if it was really possible to wait for conclusive evidence before we believe in anything?

In an essay called *The Will to Believe*, James explains that some beliefs, and our most fundamental at that, cannot possibly be proven conclusively through evidence even though we are forced to either believe in them or not. For instance, we cannot prove that it is going to rain later today and yet we have to leave the house with or without an umbrella. We have to choose either to believe that it will rain and take the umbrella, or that it won't and leave the umbrella behind - and we have to make that choice without the benefit of conclusive evidence beforehand. James called beliefs that we must make 'forced beliefs'. These forced beliefs must be decided upon one way or another, and if there is no evidence to show the way, we must decide and act with the conviction of faith.

James recognized that there are always some beliefs that are so fundamental that they must be acted upon without evidence. As he saw it, whatever we choose to believe in, or choose not to believe in, will affect the way we act and live. Therefore, how we exercise our "will to believe" is of the utmost importance. Belief and non-belief carry equal risk because human life is just as much affected by what we choose to believe in as it is by what we choose not to believe in. We stand on our beliefs and from there we push off into an uncertain future where the results of our actions will either strengthen our confidence in our beliefs or force us to reconsider them.

For James, the belief in God is an example of this type of deeply fundamental forced belief. The existence of God has been debated throughout history and it is yet to be resolved. We have no conclusive evidence either way and yet we are forced to believe in God or not, and what we decide will affect the way we live. Both the believer and the nonbeliever are ultimately standing on faith.

Some people adopt a stance of skepticism in

relation to all beliefs. They refuse to believe what is not proven for fear of believing in a lie. The activity of science largely rests in this stance and as we have already stated, it has most often served humanity well. When we talk about issues of a spiritual and moral nature however, we discover that just as often as not, conclusive evidence cannot be found. With no conclusive evidence to convince them, the skeptic decides to believe in nothing. James points out that a belief in nothing is still a belief and it still has consequences.

In opposition to the skeptical attitude, James offers a stand that he calls radical empiricism. Rather than holding back and waiting for proof, James prefers to lean forward into life, accepting that many of our decisions must be made on faith, and doing our best to choose carefully what we believe in each step of the way. Without proof we then act on our beliefs whole-heartedly as if the validity of our beliefs was assured. The process of human life is a relentless affair of jumping consciously yet somewhat blindly into the future and then continually adjusting and readjusting our beliefs based on the results of our actions. This stand in life is aligned with the pragmatic notion that the truth of an idea only reveals itself in action.

JEFF CARREIRA

"

If the hypothesis of God works satisfactorily in the widest sense of the word, it is true.

- WILLIAM JAMES

JEFF CARREIRA

09

Postmodernism and Beyond

"From the idea that the self is not given to us, I think there is only one practical consequence: we have to create ourselves as a work of art."

– MICHEL FOUCAULT

JEFF CARREIRA

THE BATTLE OF SCIENCE AND SPIRITUALITY

PRAGMATISM WAS A POWERFUL force in world philosophy during the early decades of the twentieth century. Then the First World War erupted, followed by the Great Depression, and later the horror of the Second World War. With these events, the progressive spirit of modernism, of which Pragmatism was surely a part, began to fall out of favor. Culturally, many began to believe that overemphasizing progress had resulted in a general loss of deeper human values. At Columbia where John Dewey was heading the philosophy department, he found himself intellectually opposed by a new movement, Traditionalism. Traditionalists like Mortimer Adler and Mark Van Doran were teaching their students to look toward the great works of human history to find the deep truths and spiritual values that had developed through great literature and art.

Mark Van Doran taught Literature at Columbia University and was so beloved by his students that a Mark Van Doran award was established for teaching excellence. He had a unique ability to inspire in his students a love for the human spirit as it is found in the great writings of humanity. Three of his students would go on to play significant roles in creating a new cultural movement that would challenge modernism's dominance in American life. Each was turning away from what he saw as an overly mechanistic and deterministic scientific age in search of deeper spirit. These three students were Thomas Merton, Allen Ginsberg, and Jack Kerouac. Merton would pursue the mystery of spirit as a Catholic monk and gain fame as an enormously popular author and pioneer of Interfaith dialogue through his historic meetings with Asian Spiritual Masters. Ginsberg and Kerouac plumbed the depths of their souls through poetry, prose, drugs, alcohol and Buddhism. They ignited the Beat Movement of the 1950s, which, in turn, catalyzed the hippie counter cultural movement of the 1960s. These alternative cultural movements eventually gave birth to new age spirituality and the human potential movement.

Politically and economically, the progressive side of modernity was also under attack. As the American version of democracy and the free-market economy began to show its blemishes, many turned to Socialism and Communism for answers. Philosophy, too, saw a turning away from the progressive passion of Pragmatism and any pursuit of metaphysical principles at all. Instead, philosophy embarking on a deep investigation of language and more analytical forms of philosophy. All these leanings were part of what became post-modernism, which is most simply characterized as a cultural backlash against the shortcomings of Modernism. Pragmatism was one of its victims.

Postmodernism was one anti-modernist backlash, but there was another that is interesting to consider, fundamentalism. We can see fundamentalism occurring in two forms, , religious fundamentalism and scientific fundamentalism. In the United States the sensation around the so-called Scopes Monkey Trial which challenged the rights of states to block the teaching of evolution in schools, ignited both forms of fundamentalism. Those on one side of the argument would soon argue for a literal interpretation of the bible in which God was the creator of life on Earth. On the other side scientists began to embrace a rigid view of natural selection in which chance variation accounted for all the creativity of the evolutionary process.

John Dewey was the only original Pragmatist who lived long enough to witness the decline of his life's work. He was the great progressive educational reformer who found himself the recipient of a barrage of criticism from conservative Christians, who held him personally responsible for ripping the soul out of the school system. At the same time, many scientists developed their own, at times equally fervent, scientific fundamentalist insistence that only a materialistic and deterministic interpretation of reality was scientifically valid, and the psychology of James and Dewey yielded to the growth of Behaviorism which looked at all psychological issues strictly in terms of responses to stimuli.

It now seems that we may be seeing another great turn of the cultural pendulum, as post-modern ideologies begin to show their own shortcomings. Towards the end of the twentieth century, the liberalism of the counterculture became the self-infatuation of the "Me Generation." The spiritual pursuits of the so-called new age seemed to give up reason altogether. And religious fundamentalism and scientific materialism both came under increasing scrutiny. This ignited a renewed although different spirit of progressivism, and philosophers like Richard Rorty, Hilary Putnam and Cornel West began to dust off and reinterpret Pragmatism for a new era. Even more recently, a pioneering group of pioneering thinkers referring to themselves as Speculative Realists are reviving metaphysical speculations. These philosophers, often motivated by our world's mounting challenges, believe that we must radically revision our understanding of reality and to do that we must encourage, not discourage, speculation about the nature of reality.

Perhaps it is time for philosophy to take up a central position in the public's eye once again. In America, Ralph Waldo Emerson, William James, and John Dewey were all enormously popular and respected public figures. Margaret Fuller was famed for holding public gatherings called "conversations," early progenitors of feminist consciousness raising events. Jane Addams, a close colleague of John Dewey, was the second woman ever to receive a Noble Peace Prize. Today philosophy enjoys very little public attention, but it may be a time when we need to capture people's imagination with exciting ideas that can transform society. In a world as tumultuous as the one in which we live, the need for deeper philosophical introspection could not be greater.

Only the dreamer shall understand realities, though in truth his dreaming must be not out of proportion to his waking.

- MARGARET FULLER

JEFF CARREIRA

10

Final Thoughts and Conclusions

"Truth can be stated in a thousand different ways, yet each one can be true."

– SWAMI VIVEKANANDA

JEFF CARREIRA

THE BATTLE OF SCIENCE AND SPIRITUALITY

ONE OF THE FIRST books I wrote was called *Philosophy Is Not A Luxury* because to me philosophy is an essential occupation of human life. We all have to make decisions and choices and those decisions will always rest on some rationale. There are reasons why we do the things we do. What we believe about ourselves, about life, and about reality, inevitably affect how we live. We are philosophical beings whether we realize it or not. I hope that this book has given you some things to think about as you make the choices that will define your life.

I also hope that this quick run through some of the history of Western philosophy has explained what many of us see today as the deep polarizations that exist among human beings. People often line up on one side of a debate or the other. There are too many domains in which polarization can be found to possibly mention them all, but certainly the line that divides those that are politically conservative versus liberal comes to mind, as does the line that has been central to the discussion of this book, the division between science and spirituality.

These pages have offered you a very brief exploration of some turning points in the Western philosophical tradition to give you some insight into the origin of some of the most fundamental divisions that separate us. Early on we distinguished between the rational temperament from the romantic one. We explained the separation that divides those that are spiritually inspired from those with a scientific orientation. The origin of these dividing lines may be buried as deeply in the human experience as the division between mind and body and the distinction between our thoughts and our feelings. It has also been speculated that it lies in the different ways that the two halves of our brains operate. Whatever the ultimate origin, it seems that in the Western philosophical tradition our earliest cultural beliefs were rooted in two different, and often conflicting, understandings of truth and how to find it. It is my hope that a deeper historical understanding of the polarizing influence that have shaped our mental attitudes will help us transcend and heal the divisions that exist between us.

Three things cannot be long hidden: the sun, the moon, and the truth.

~ BUDDHA

JEFF CARREIRA

About the Author

Jeff Carreira is a meditation teacher, mystical philosopher and author who teaches to a growing number of people throughout the world. As a teacher, Jeff offers retreats and courses guiding individuals in a form of meditation he refers to as The Art of Conscious Contentment. Through this simple and effective meditation technique, Jeff has led thousands of people in the journey beyond the confines of fear and self-concern into the expansive liberated awareness that is our true home.

Ultimately, Jeff is interested in defining a new way of being in the world that will move us from our current paradigm of separation and isolation into an emerging paradigm of unity and wholeness. He is exploring some of the most revolutionary ideas and systems of thought in the domains of spirituality, consciousness, and human development. He teaches people how to question their own experience so deeply that previously held assumptions about the nature of reality fall away to create space for dramatic shifts in understanding.

Jeff is passionate about philosophy because he is passionate about the power of ideas to shape how we perceive reality and how we live together. His enthusiasm for learning is infectious, and he enjoys addressing student groups and inspiring them to develop their own powers of inquiry. He has taught students at colleges and universities throughout the world.

Jeff is the author of numerous books including:

The Art of Conscious Contentment, No Place But Home, The Miracle of Meditation, The Practice of No Problem, Embrace All That You Are, Philosophy Is Not a Luxury, Radical Inclusivity, The Soul of a New Self, and *Paradigm Shifting.*

For more about Jeff or to book him for a speaking engagement, visit: jeffcarreira.com

THE BATTLE OF SCIENCE AND SPIRITUALITY

Made in United States
North Haven, CT
17 June 2024

53721536R00065